Taking a Camera To Germany

Ted Park

ELEMENTARY · SECONDARY · ADULT · LIBRARY

A Harcourt Company

www.steck-vaughn.com

Copyright © 2001, Steck-Vaughn Company

ISBN 0-7398-4130-0

All rights reserved. No part of this book may be reproduced or utilized in any form or by any means, electronic or mechanical, including photocopying, recording, or by any information storage and retrieval system, without permission in writing from the publisher. Inquiries should be addressed to copyright permissions, Steck-Vaughn Company, P.O. Box 26015, Austin, TX 78755.

Printed and bound in the United States of America
10 9 8 7 6 5 4 3 2 1 W 05 04 03 02 01

Cover photo: Brandenburg Gate

Photo acknowledgments

Contents

You're in Germany!	4
Looking at the Land	6
Berlin	10
Great Places to Visit	12
The People	14
How Do People Live in Germany?	16
Government and Religion	18
Earning a Living	20
School and Sports	22
Food and Holiday Fun	24
The Future	26
Quick Facts About Germany	28
Glossary	30
Index	32

You're in Germany!

Germany is a country in the center of Europe. Germany has a mix of mountains, forests, flat lands, and gentle hills. There are beaches in the north, and mountains for skiing in the south.

The German countryside is dotted with ancient churches, cathedrals, and castles. A cathedral is a large church. The **Black Forest** is in the south and is one of Germany's most famous places. Germans have made cuckoo clocks in the Black Forest for hundreds of years. Take a boat ride down the **Rhine River**. The river's valley has beautiful little towns and bridges that would make great pictures.

Germany has five large cities: **Berlin**, **Hamburg**, **Munich**, **Cologne**, and **Frankfurt**. Berlin is the largest city and the capital. Germany's main seaport is in Hamburg. Munich is the capital of the Bavarian region. Cologne is famous for its cathedral. Frankfurt is the financial center of Germany.

The famous Cathedral of Cologne overlooks the Rhine River. Construction of the cathedral started in 1248 and was completed in 1880.

This book will show you some of the best things to see in Germany. You will learn interesting things about the country and the people who live there. So, when you're ready to take your camera there, you'll know exactly what to do and where to go. Enjoy your trip!

 # Looking at the Land

Denmark and the Baltic and North Seas border Germany in the North. France, Belgium, the Netherlands, and Luxembourg border Germany on the west, and Poland and the Czech Republic are on the east. Switzerland, Liechtenstein, and Austria border Germany to the south.

Germany is about 520 miles (840 km) from north to south and about 385 miles (620 km) from east to west. Germany has 137,803 square miles (356,910 sq km) of land within its borders. This means Germany is about the size of the state of Montana.

Denmark is a peninsula that extends from northern Germany. A peninsula is a piece of land that has water on three sides. The North and the Baltic Seas surround Denmark. Germany has its main seaports on the North Sea. The Baltic Sea has islands and beaches where people go for vacations.

People take tours along the Rhine River to see its beautiful buildings.

Mountains and Plains

The country is hilly in the western and central parts. Southern Germany has snowy mountains and thick forests. The farther south you travel, the taller the mountains get. These tall mountains are known as the **Bavarian Alps**. The **Zugspitze** is Germany's highest alpine peak. It is 9,721 feet (2,963 m) high.

Germany has many rivers. The most famous one is the Rhine River. Other rivers include the Elbe, Weser, Ems, and Main. Most of Germany's rivers are used to transport things the German people need or sell to others. Transport means to move something from one place to another.

Germany generally has moderate or mild weather. This means it is usually never too hot or too cold. Even in the highest mountains, warm winds keep the weather mild.

Bavaria, located in Germany's southern region, has many rich, green forests and snowy mountains.

Berlin

Germany's Capital

Four years after World War II ended, Germany was divided into two countries: East Germany and West Germany, and the city of Berlin was divided into two cities. The Soviet Union, now called Russia, ruled East Berlin and East Germany. The Americans, French, and British ruled the west. During this time, West Germany's capital was the city of Bonn. The city of West Berlin was like a small island of freedom where people could escape communism. Under communism, people work for and are taken care of by the state.

In 1961, the Soviets put up the famous **Berlin Wall**. The wall was built to keep East Germans from escaping to West Berlin. The East Germans would try to escape so that they could live in a free country. The Berlin Wall was taken down in 1989 and Germany was made into one country again. Then the capital was moved from Bonn back to Berlin. During the 1990s,

There are museums, palaces, and historic buildings on the same street as the famous Brandenburg Gate.

much of Berlin was rebuilt. The **Reichstag** or parliament building was rebuilt and became the center of Germany's government again.

Today Berlin is new and exciting. The **Brandenburg Gate** is one of Berlin's most famous sights. It is at the beginning of Unter den Linden, a wide, tree-lined street. If you visit Berlin, remember to take a picture of this beautiful building.

 # Great Places to Visit

The first place you should visit is Hamburg. It is Germany's main seaport and second largest city. Streams and canals cross through the city. Canals are waterways that help river traffic. Hamburg was bombed during World War II and a lot of the city was destroyed. Today, much of the city has been rebuilt. Although Hamburg has a great deal of industry, it also has parks and gardens.

Make sure you stop in Munich. It is the third largest city. Munich is the capital of the region of Bavaria. There is a famous palace in the middle of the city. This palace once belonged to the kings of Bavaria. Now it is a museum.

Don't forget that the Rhine Valley is one of Germany's most beautiful places. The river winds through the country. Nearby there are vineyards, where grapes are grown, and from the river, you will see beautiful old castles.

This old castle has a great view of the Rhine Valley.

# The People

More than 82 million people live in Germany. This makes Germany one of the most crowded countries in Europe. About a third of all Germans live in cities, mainly in the western part.

Depending on where they live, Germans may look very different. In the south, in Bavaria, many people wear traditional clothing such as **lederhosen**. Lederhosen are leather shorts with suspenders. This is clothing that people in this region have worn for many years.

Germany did not become one country until 1871, therefore, people are more loyal to the region they live in than to the country. Many people in Bavaria still think of themselves as Bavarians first and Germans second.

Today, there are people in Germany who have come from other places looking for work. They are known as **gastarbeiter**, which means guest worker. These people came from eastern Europe, Italy, and Turkey.

◀ This man wears a traditional costume while playing a tune on his accordion.

People who live ▶ in Bavaria wear the traditional dress of that region.

Today, almost one tenth of the people living in Germany were born in another country.

High German is the language that developed in the regions of the south, but Germans speak differently in other parts of the country. These differences are known as **dialects**. A dialect is a different way of speaking the same language.

How Do People Live in Germany?

Germans work hard but they have a lot of time to play. Most Germans get four to six weeks of vacation time each year. They like to travel. They like outdoor activities, and they often visit spas, which are places where people can get exercise and fresh air.

Germans are book lovers and readers. In fact, after the United States, Germany has the highest number of books published. Published means any printed material such as a book that is produced so that people can buy it. Germans also like to visit museums.

For breakfast, they have a type of meat, cheese, and bread. Their biggest meal is at lunchtime and dinner is smaller.

Germany has many new and old buildings side-by-side. ▶

Government and Religion

Government

Germany is made up of 16 states, called **Länder**. The states are joined together, which makes Germany a federal republic. The German parliament makes the laws. It has two chambers, or branches. They are the Federal Assembly and the National Assembly. The head of the government is known as the **chancellor**. He or she is the leader of the party that has been elected. Elect means to choose someone by voting. There is also a president, but that person has less power than the chancellor.

Religion

About a third of all Germans are Protestant. Most Protestants live in the north, and most of them belong to the Lutheran church. Germany was the home of Martin Luther. He broke away from the Catholic Church and founded the Protestant religion.

While in Germany, don't forget to take a walk by Old Town Hall.

Another third of all Germans are Roman Catholics. Most Catholics live in the southern part of the country. There are also about a million Muslims in Germany. Most of the German Muslims came from the country of Turkey.

Earning a Living

Germany is a country rich in **natural resources**. A natural resource is something that is found in nature and is necessary or useful to people. Coal and iron ore are just two of these resources. Many people work as miners to find them.

About 33 percent of Germans work in industry. They are skilled at turning natural resources into man-made products. Most people work for small companies. Some of the smallest companies make the most well-known crafts. In the north, making ships is an important industry. Chemicals, steel, medicines, cars, and cameras are among the larger industries.

The automobile was invented in Germany in the 1870s. The car industry is Germany's biggest. The Volkswagen was first built in Germany. It is also the home of the Mercedes Benz and BMW cars.

About half the land in Germany is used for farming, but only a small number of people are farmers.

◀ The automobile was invented in Germany in the 1870s. Old-fashioned machines used to work on cars. Now, these robots do the job.

Germany's main ▶ seaport is on the North Sea.

School and Sports

School

Most German children attend kindergarten between the ages of 4 and 6. Kindergarten is a German word that means *children's garden* in English. Children attend primary school from ages 6 to 10, after which children split into three categories: trade or vocational school, business school, or college preparatory schools. College preparatory schools usually take nine years to complete. In many areas, children attend Saturday classes.

Germany has many universities, where students learn to become scientists and engineers, among many other professions. Germany's oldest university is at Heidelberg, which was founded in 1386.

Sports

Germans like to spend time outdoors. They enjoy skiing, camping, hiking, and bicycling. Soccer is the most popular sport. Germans like to play soccer and to watch favorite teams play. Many places sponsor sports

German students study hard. Some become doctors.

clubs. Some large German cities have American-style football and basketball teams. German athletes often win a large share of the medals at the Olympics games.

Food and Holiday Fun

Let's Eat!

Germany is famous for its rich food. Many kinds of sausages, breads, potatoes, dumplings, sauerkraut, and of course frankfurters are German staples. Desserts include cakes, gingerbread, strudel, and stollen, a rich coffeecake.

Each year, in Munich, there is **Oktoberfest**. This is a fall harvest festival, where food and drink are the most important things.

Celebrate!

Many German cities have special markets in the weeks before Christmas. Shoppers can buy all kinds of food, decorations, toys, and other gifts. Germans also celebrate Advent, the four weeks before Christmas. Advent calendars and Christmas trees come from Germany.

The 40-day period just before Easter is known as

The idea of the Christmas tree came from Germany. If you go there during this time of year, don't forget your camera!

Lent. Fasching, or **Karneval** (this is how the German people spell **Carnival**), comes just before Lent starts. It is a time for parties, parades, and dressing up in costumes.

The Future

If you took your camera to Germany, you would see a country that is changing quickly. During the 50 years that Germany was divided, the western part of the country developed quickly. The eastern part fell behind. Today, Germany is trying to help the former East Germany catch up.

However, Germany has one of the world's strongest economies. Many large companies have their main offices in Germany. The former East Germany, however, still needs a lot of help. A great deal of money is being spent to improve highways and railroads in the east. Germans are spending money to reduce noise and air pollution. People are trying to protect the forests for the future.

Germans are proud of their country. When you leave Germany, people will say Auf Wiedersehen (owf VEE-dare-zane). These are the German words that mean good-bye in English.

**You must visit Germany's parliament building, the Reichstag. ▶
Have a great trip!**

Quick Facts About Germany

Capital
Berlin

Borders
Denmark, Netherlands, Belgium, Luxembourg, France, Switzerland, Austria, Czech Republic, and Poland

Area
137,803 square miles
(356,910 sq km)

Population
82,087,361

Largest Cities
Berlin (3,458,763 people)
Hamburg (1,707,986 people)
Munich (1,225,809 people)
Cologne (964,346 people)
Frankfurt (647,304 people)

Chief crops
grains, potatoes, sugar, beets

Natural resources
coal, potash, lignite, iron, uranium

Chief rivers
Elbe, Weser, Ems, Rhine, and Main

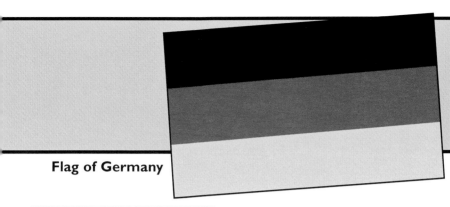

Flag of Germany

Coastline
1,493 miles (2,389 km)

Monetary unit
mark

Literacy rate
100%

◀ **Major industries**
steel, ship building, vehicles, machinery, electronics, coal, chemicals, iron, cement, food, and beverages

Glossary

Bavarian Alps: (Buh-VARE-ee-in ALPSS) The tall mountains in the south of Germany.

Berlin Wall: (bur-LIN WAWL) A wall that was built to keep East Germans from escaping to West Germany.

Berlin: (bur-LIN) The capital of Germany, and its largest city.

Black Forest: (BLAK FOR-ist) A famous forest in Germany where cuckoo clocks have been made for hundreds of years.

Brandenburg Gate: (BRAN-den-burg GATE) A gate that was built from 1788-1791 as an entrance to the city of Berlin.

Chancellor: (CHAN-suh-lur) The head of the government in Germany.

Cologne: (kuh-LONE) The fourth largest city in Germany that has 963,300 people.

Dialect: (DYE-uh-lekt) A way a language is spoken in a particular place or among a group of people.

Fasching: (FOSH-ing) This is a carnival that takes place just before the start of Lent. It is also called Karneval.

Frankfurt: (FRANK-furt) Germany's fifth largest city. It has the country's largest airport and biggest university.

gastarbeiter: (GAHST-a-bate-uh) A person who has

come to Germany to look for work. This word means guest worker.

Hamburg: (HAM-burg) The second largest city in Germany, and its main seaport.

High German: (HYE GUR-men) The language that developed in the regions of southern Germany.

Karneval: (KAR-nuh-vuhl) This is a celebration, also known as Fasching, that happens just before Lent starts.

länder: (LEN-duhr) The 16 states that make up Germany.

lederhosen: (LA-dur-hoh-zen) Leather shorts that are held up with suspenders.

Lent: (LENT) The seven-week period leading up to Easter.

Munich: (MEW-nik) The capital of the Bavarian region and Germany's third largest city.

natural resource: (NACH-ur-ruhl REE-sorss) Something that is found in nature that is necessary or useful to people.

Oktoberfest: (ok-TOH-bur-fest) A fall harvest festival, where food and drink are important.

Reichstag: (RIKE-stog) A building that is the center of Germany's government. Reichstag means parliament.

Rhine River: (RINE RIV-ur) A river that winds through most of Germany.

Zugspitze: (ZOOG-shpit-zuh) The highest Alpine peak in Germany.

Index

Alps, Bavarian 8
automobile 20

Bavaria 12, 14
Berlin 4, 10-11
Berlin Wall 10
Black Forest, The 4
Bonn 10, 11
Brandenburg Gate 11

dialects 15

East Germany 10, 26
Eastern Europe 14

Fasching (Karneval) 25

Gastarbeiter (guest worker) 14

Hamburg 4, 12
Heidelberg University 22
High German 15

Italy 14

Länder (states) 18
language 15
lederhosen 14

Main River 4
Munich 4, 12, 24

Oktoberfest 24

parliament 11

Reichstag 11
Rhine River 8
Rhine Valley 12

seaports 6
Soviet Union 10-11
spas 16

Turkey 14, 19

Unter den Linden 11

weather 9
West Germany 10
World War II 10, 12

Zugspitze 8